**PEARSON**

# Math
## Makes Sense

Phoenix
320 19
Calgary AB T2E 6J6

## Author Team

Carole Saundry

Maureen Dockendorf

Michelle Jackson

Maggie Martin Connell

Sharon Jeroski

Cathy Anderson

Brenda Lightburn

Michelle Skene

**PEARSON**

Education
Canada

**Publisher**
Claire Burnett

**Publishing Team**
Enid Haley
Lesley Haynes
Cristina Getson
Ellen Davidson
Lynne Gulliver
Stephanie Cox
Jane Schell
Judy Wilson

**Design and Art Direction**
Word & Image Design Studio Inc.
Carolyn Sebestyen

**Math Team Leader**
Diane Wyman

**Product Manager**
Kathleen Crosbie

ISBN-13:  978-0-321-41734-3
ISBN-10:   0-321-41734-8

Printed and bound in the United States

12 13 14 15  CC  15 14 13

The information and activities presented in this
book have been carefully edited and reviewed.
However, the publisher shall not be liable for any
damages resulting, in whole or in part, from the
reader's use of this material.

The publisher has taken every care to meet or
exceed industry specifications for the
manufacturing of textbooks.

Brand names that appear in this book are intended
to provide children with a sense of the real-world
applications of mathematics and are in no way
intended to endorse specific products.

**Acknowledgments**
The publisher wishes to thank the following sources
for photographs, illustrations, and other materials
used in this text. Care has been taken to determine
and locate ownership of copyright material in this
book. We will gladly receive information enabling us
to rectify any errors or omissions in credits.

**Cover**
Cover illustration by Marisol Sarrazin © 2001. Taken
from *Nose to Toes* by Marilyn Baillie with permission
of Maple Tree Press Inc.

**Illustrations**
Virginie Faucher, pp. 24, 139–147, 148 (licence plate,
pizza slice, bean), 149, 150, 155, 156; Leanne Franson,
pp. 41–54; Linda Hendry, pp. 14, 15, 18–23, 25, 27–33,
36–40, 56, 85–93, 94 (desktop, calendar), 95, 102,
113, 153, 157, 160; Tina Holdcroft, pp. 16, 17, 96–100,
148 (button, die), 173–184; Vesna Krstanovic, pp. 1–4,
6–12, 26, 35, 68–70, 94 (boy with paper), 103–112,
114, 118, 121–125, 134, 135, 154, 158, 159, 162, 163,
167, 169, 171, 172; Isabelle Langevin, p. 13; Albert
Molnar, pp. 59–67, 71–77, 80–84; Wolf Morrisseau,
Ojibway, Garfinkel Publications, Vancouver, p. 161
(*Moose; Bear*); Photodisc/Getty Images, p. 101; Michel
Rabagliati, p. 116 (scale); Anne Villeneuve, pp. 119,
120, 126–133, 136–138, 164, 165, 168, 170; Mervin
Windsor, Haisla, Heiltsuk, Garfinkel Publications,
Vancouver, p. 161 (*Salmon*)

**Photography**
Gilbert Duclos, pp. 55 (girl with helmet), 115 (boy
holding onto bar), 185 (girl planting seeds); Ian Crysler,
pp. 55 (children searching), 57 (children with play
dough; coins; children and coins), 115 (girl;
containers), 116 (girl and mirror), 117, 118, 185 (rapper
girl), 186 (groceries), 187, 188; Ray Boudreau, pp. 56, 57
(crayons, pattern items), 58, 186 (boys and cars)

# Contents

## Advisers

### Consultants

Craig Featherstone
Maggie Martin Connell
Trevor Brown

*Assessment Consultant*
Sharon Jeroski

*Elementary Mathematics Adviser*
John A. Van de Walle

*British Columbia Early Numeracy Project Contributor*
Carole Saundry

*Cultural Adviser*
Susan Beaudin

### Advisers

Pearson Education thanks its Advisers, who helped shape the vision for *Pearson Math Makes Sense* through discussions and reviews of prototype materials and manuscript.

| | | | |
|---|---|---|---|
| Joanne Adomeit | Brenda Foster | Jodi Mackie | Cheryl Shields |
| Bob Berglind | Marc Garneau | Ralph Mason | Gay Sul |
| Auriana Burns | Angie Harding | Christine Ottawa | Chris Van Bergeyk |
| Edward Doolittle | Florence Glanfield | Gretha Pallen | Denise Vuignier |

# Patterning

 **FOCUS**

Children talk about the patterns in the picture.

 **HOME CONNECTION**

Ask your child to look at the picture and describe the patterns.

Name: _____     Date: _____

Dear Family,

Your child is starting the first unit in mathematics and will be learning about patterning.

The Learning Goals for this unit are to

- Identify and copy patterns.
- Create and extend patterns.
- Talk about how a pattern repeats.
- Solve problems with patterns.
- Recreate the same pattern using different materials.

You can help your child reach these goals by doing the activities suggested at the bottom of each page.

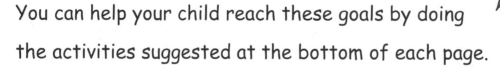

# Copy a Pattern

Use cubes to make a pattern.
Copy the pattern.

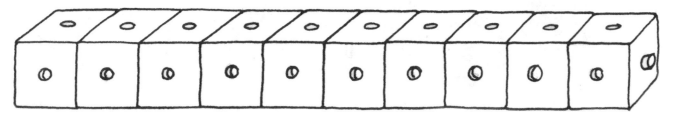

Make another pattern.
Copy the pattern.

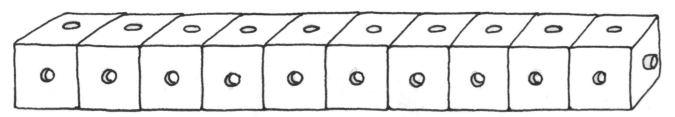

Choose a friend's pattern.
Copy the pattern.

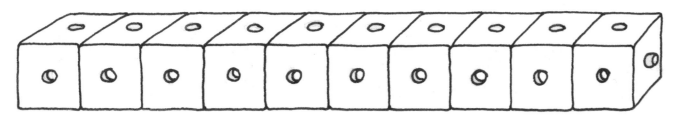

 **FOCUS**

Children use Snap Cubes or blocks to make a
pattern. Then they copy the pattern by colouring the
blocks on the page.

 **HOME CONNECTION**

Go on a pattern hunt at home with your child. See
how many things you can find that have patterns
(sweaters, socks, rugs, dishes). Ask your child to
describe the patterns.

**Unit I, Lesson I:** Recognize and Copy a Pattern   **3**

Name: _____  Date: _____

# Colour Patterns

Colour to show a pattern.

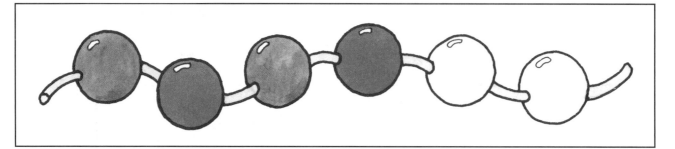

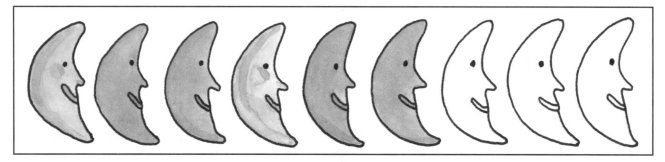

# Make a Pattern

Make 2 different patterns.

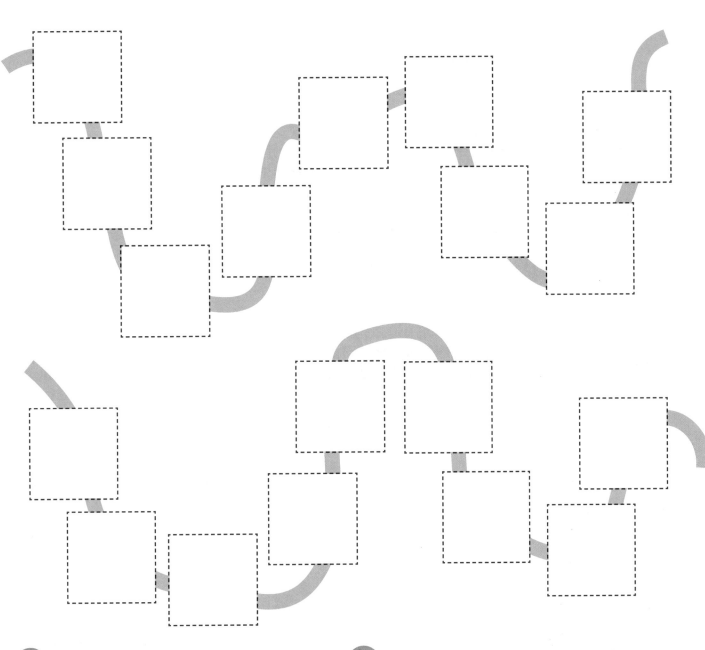

 **HOME CONNECTION**

With your child, cut out pictures from newspaper flyers or magazines to make a pattern (vegetable, vegetable, fruit).

**Unit I, Lesson 2:** Make and Extend a Pattern **5**

Name: _____  Date: _____

# What Comes Next?

Draw what comes next in the pattern.

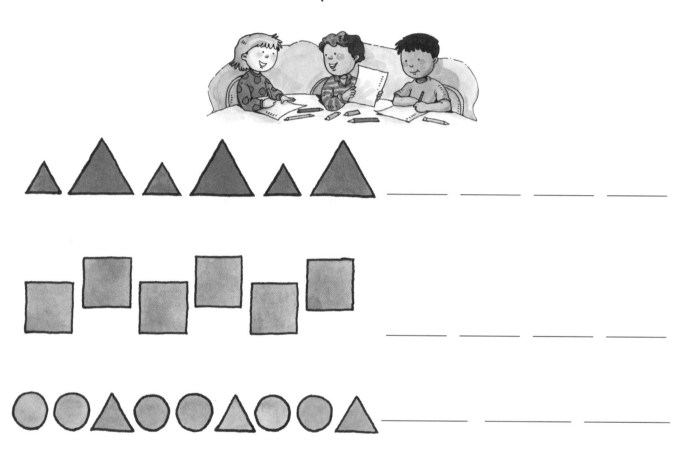

Make a pattern. Ask a friend to draw what comes next.

<br>
<br>
<br>
<br>
<br>

 **FOCUS**

Children name and extend a pattern.

 **HOME CONNECTION**

With your child, use common items like plastic juice-bottle caps or stickers to begin a pattern. Then, ask your child, "What comes next in the pattern?" Complete the pattern together.

Name: _____ Date: _____

# Pattern Practice

Make a colour pattern.

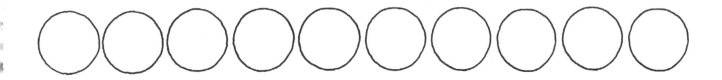

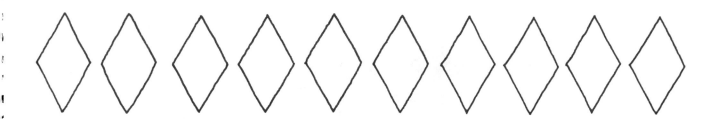

**FOCUS**

Children make and extend a pattern.

**HOME CONNECTION**

Use household objects to practise making many different kinds of patterns (spoon, spoon, fork; spoon, spoon, fork; spoon, spoon, fork).

**Unit I, Lesson 2:** Make and Extend a Pattern    **7**

Name: _____    Date: _____

# Find the Pattern

Some beads fell off a necklace. What colour are they?
Colour the missing beads.

**FOCUS**

Children identify the pattern and colour the missing beads.

**HOME CONNECTION**

With your child, begin a pattern such as mug, mug, glass; mug, mug, glass. Ask your child to complete the pattern.

**8**   **Unit I, Lesson 3:** Strategies Toolkit

Name: _____  Date: _____

# Fill in the Pattern

Find the pattern. Fill in ☐ to show what is missing.

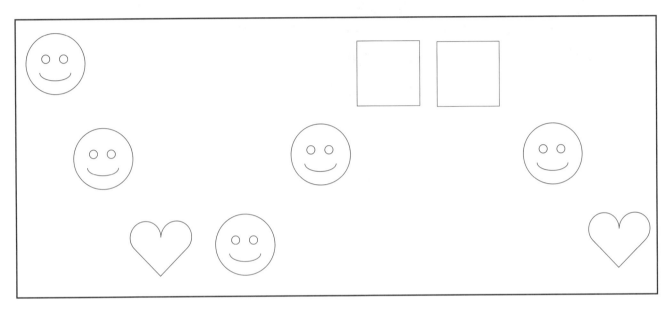

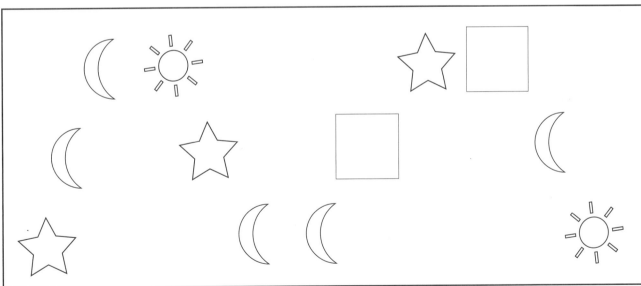

**FOCUS**

Children draw to identify the missing parts of each pattern.

**HOME CONNECTION**

With your child, make a pattern with pens, pencils, crayons, or markers. Repeat the pattern three times. Remove two objects. Ask your child: "What is missing? How do you know?"

Name: _____  Date: _____

# Show the Same Pattern

Look at the Snap Cube Pattern.

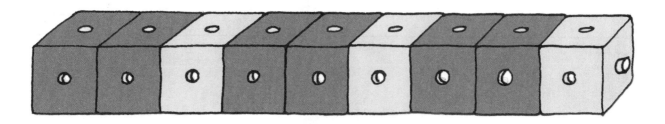

Show the pattern 2 other ways.

Name: _____ Date: _____

# My Pattern Border

Draw the pattern for your border.

Name: _____  Date: _____

# My Journal

Tell what you learned about patterning.
Use pictures or words.

**FOCUS**

Children reflect on and record what they learned about patterning in this unit.

**HOME CONNECTION**

Ask your child: "What pattern could you make using your favourite people, toys, or books?"

# UNIT 2

# Representing Numbers to 20

 **FOCUS**

Children talk about the picture and identify the numbers of objects.

 **HOME CONNECTION**

Together, look at the picture and ask: "Where is there a group of four? a group of six? How do you know?" Have your child point to each object while counting aloud.

Name: _____ Date: _____

Dear Family,

This unit will focus on deepening your child's understanding of number relationships.

The Learning Goals for this unit are to

- Read and print numbers to 20.
- Read number words to twenty.
- Count forward from 0 to 20. Count backward from 20 to 0.
- Count by matching the number word to the objects being counted.
- "Build" numbers by arranging and rearranging objects.
- Recognize familiar groups of objects or dots.
- Compare and order numbers and groups of objects using words such as *more, fewer,* and *same.*
- Estimate the number of objects and check by counting.
- Use real-life materials to help solve simple number problems.

You can help your child reach these goals by doing the activities suggested at the bottom of each page.

Name: _____  Date: _____

# Pictures of 6

Use counters to show 6.

Draw pictures of ways you can show 6.

 **FOCUS**

Children show the number 6 in a variety of ways.

 **HOME CONNECTION**

Use buttons, keys, or other small objects and ask your child to show the number 4 in different ways.

Name: _____     Date: _____

# Ten Frame Numbers

**My number:** _____

### Draw counters to show I more:

|  |  |  |  |  |
|---|---|---|---|---|
|  |  |  |  |  |

### Show I less than your number:

|  |  |  |  |  |
|---|---|---|---|---|
|  |  |  |  |  |

### Show 2 more than your number:

|  |  |  |  |  |
|---|---|---|---|---|
|  |  |  |  |  |

### Show 2 less than your number:

|  |  |  |  |  |
|---|---|---|---|---|
|  |  |  |  |  |

---

**My number:** _____

### Draw counters to show I more:

|  |  |  |  |  |
|---|---|---|---|---|
|  |  |  |  |  |

### Show I less than your number:

|  |  |  |  |  |
|---|---|---|---|---|
|  |  |  |  |  |

### Show 2 more than your number:

|  |  |  |  |  |
|---|---|---|---|---|
|  |  |  |  |  |

### Show 2 less than your number:

|  |  |  |  |  |
|---|---|---|---|---|
|  |  |  |  |  |

**FOCUS**

Children draw counters on ten frames to represent I more, I less, 2 more, and 2 less than a given number.

**HOME CONNECTION**

Make a ten frame on paper or by cutting 2 sections off an egg carton. Suggest a number and have your child use objects and the ten frame to show you I less or I more than the number.

Name: _____  Date: _____

# Terrific Ten!

Count the number of counters.
Record the numbers.

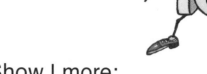

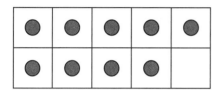

## Show 1 more:

_____

_____

## Show 2 less:

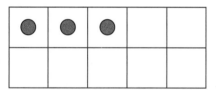

_____

_____

(My number) Draw counters to show your number.

## Show 2 more:

Record the number:

Record the number:

_____

_____

# Flash It!

Choose a number. Draw a flash card representing your number.

| | | | | |
|---|---|---|---|---|
| | | | | |
| | | | | |

Choose a different number. Draw a flash card representing your new number.

| | | | | |
|---|---|---|---|---|
| | | | | |
| | | | | |

**FOCUS**

Children draw their own flash cards on ten frames.

**HOME CONNECTION**

Invite your child to tell you how to play Flash It, then play Flash It using a deck of playing cards (2s to 10s only). Cover the numerals with small self-stick notes. Have your child check her or his answers by counting.

Name: _____  Date: _____

# Show Numbers to 20

Choose a number between 10 and 20.
Draw counters to show the number on the two-part mat.

My number _____

| | |
|---|---|
| | |

Choose another number. Draw counters to show it.
My number _____

| | |
|---|---|
| | |

**FOCUS**
Children identify, record, and represent numbers to 20 on a two-part mat.

**HOME CONNECTION**
Ask your child to show the numbers on this page in a different arrangement.

Name: _____     Date: _____

# How Many Apples?

There are 13 apples in a bag.
Some are yellow. Some are red.
How many of each could there be?

Show your thinking in pictures, numbers, or words.

**FOCUS**

Children figure out possible combinations of 13 red and yellow apples. They express their solutions using pictures, numbers, or words.

**HOME CONNECTION**

When problem solving, your child may discover more than one correct answer. Ask your child to explain how he or she solved this problem.

**28**  Unit 2, Lesson 7: Strategies Toolkit

# What Is in the Backpack?

There are 11 things in the backpack.
Some are books. Some are snacks.

How many could be books? _____

How many could be snacks? _____

Show your thinking in pictures, numbers, or words.

**FOCUS**

Children figure out possible combinations of 11 books and snacks. They express their solutions using pictures, numbers, or words.

**HOME CONNECTION**

Tell your child: "I have 11 spoons and forks. How many of each might I have?" Give your child spoons and forks to find out the different combinations to make 11.

# Ways to Make 15

Show different ways to make 15.
Make equal groups without singles.
Make equal groups with singles.

 **FOCUS**

Children show different ways to make 15,
using equal groups with and without singles.

**HOME CONNECTION**

Together, find 15 small objects in your house and
separate them into equal groups in as many ways
as possible.

**30**   **Unit 2, Lesson 8:** Grouping Numbers to 20

Copyright © 2007 Pearson Education Canada **Not to be copied.**

Name: _____  Date: _____

# Build Numbers

Pick a number from 11 to 19. Write your number in the middle box.
Build your number in the middle set of ten frames.
Build the numbers that come before and after your number.

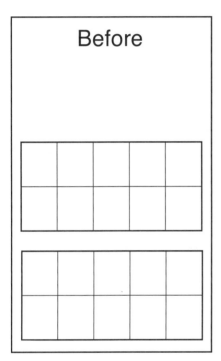
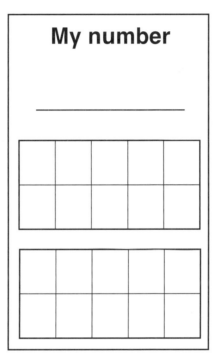
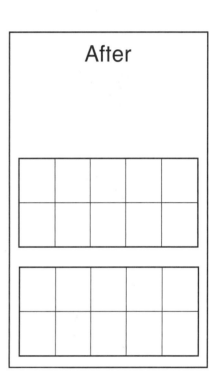

| Before | My number | After |
| --- | --- | --- |
| | _____ | |

How did you know what to build?
Tell a friend.

**FOCUS**

Children "build" a chosen number using ten frames, then build the numbers that come before and after the chosen number.

**HOME CONNECTION**

Make 3 ten frames on paper or use sections of egg cartons. Ask your child to repeat the activity on this page using a different middle number.

Name: _____  Date: _____

# About How Many?

Estimate the number of cubes.
Spill and count the cubes.

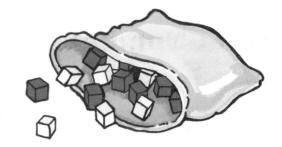

| Cubes | Estimate | Count |
|-------|----------|-------|
|  | about | |
| | about | |
| | about | |

There are _____  and _____  .

There are _____ cubes in all.

Name: _____  Date: _____

# Favourite Activity

Ask 6 friends this question:

Of swimming and bike-riding, which do you like better?

Record your friends' choice.

| Swimming | Bike-riding |
|----------|-------------|
|          |             |
|          |             |
|          |             |
|          |             |

Which is the favourite activity? _____

Compare the groups. Use pictures, numbers, or words.

|   |
|---|
|   |

**FOCUS**
Children compare their friends' favourite activities.

**HOME CONNECTION**
Have your child ask family members the same question from the activity, then compare the groups.

# More or Less

Draw counters to show more or less.

| | |
|---|---|
| Show 2 more. | |
| Show 1 more. | |
| Show 1 less. | |
| Show 2 less. | |
| Show 1 more. | |
| Show 2 more. | |

**FOCUS**
Children represent numbers by drawing counters that show one more, one less, two more, two less.

**HOME CONNECTION**
Ask your child to describe and explain the number of counters he or she drew.

Name: _____  Date: _____

# Number Challenge

Print the numbers.
Here are your clues.

One more than eight                    _____

Two more than five                     _____

One less than three                    _____

Two less than six                      _____

One more than four                     _____

Two less than five                     _____

Two more than four                     _____

# What You Know about 17

Build 17 in two ways.
Use pictures, numbers, or words.

| 17  seventeen |
| --- |
| |

| 17  seventeen |
| --- |
| |

What other ways can you show 17? Tell a friend.

 **FOCUS**
Children use objects to build 17 in different ways.
They record their thinking, using pictures, numbers,
or words.

 **HOME CONNECTION**
This activity gives your child a chance to show what
he or she has learned about numbers. Ask: "How
did you know so many different ways to build 17? Tell
me about your thinking."

Name: _____  Date: _____

# 20! 20! 20!

Show 20 objects in two groups.
Show as many different ways as you can.

Name: _____  Date: _____

# My Journal

Tell what you learned about numbers.
Use pictures, numbers, or words.

---

 **FOCUS**

Children reflect on and record what they learned
about numbers to 20.

**HOME CONNECTION**

Invite your child to share thoughts about working with
numbers in this unit. Ask: "What have you learned
about numbers to 20?"

**Unit 2, Lesson 13:** Show What You Know

# Mystery Bags
# for School

**Take-Home Story**

I'm making mystery bags for fun.
I need some things for everyone.

I can choose things that are round,
like coins or buttons I have found.

I can choose things that are square,
like stamps or cubes that I can share.

I can choose things red or blue, purple,
yellow, orange, too.

I can choose things big or small,
like marbles or this red-striped ball.

I'll count these things, then I'll be done.
Mystery bags are lots of fun!

# From the Library

Ask the librarian about other good books to share about patterning and numbers.

Eric Carle, *The Grouchy Ladybug* (HarperCollins, 1996)

Phoebe Gilman, *Jillian Jiggs* (Scholastic, 1988)

Tana Hoban, *Let's Count* (Greenwillow Books, 1999)

Bill Martin, Jr., *Brown Bear, Brown Bear, What Do You See?* (Henry Holt and Company, 1992)

Jan Thornhill, *The Wildlife 1-2-3: A Nature Counting Book* (Greey de Pencier, 2003)

# Solve the Mystery!

Which bag is Jada's?

How can we help Miss Flish find out?

# What's in Your Mystery Bag?

Look in your bag.
Do you think there are more than 10,
or fewer than 10 objects? _____
What's inside? Draw what you have.

How many objects are in your bag? _____

# More, Fewer, or As Many?

Draw your objects in 2 groups.

```



```

Do you have more, fewer, or as many objects as the
special mystery bag? _____
How do you know?

```



```

# Make a Pattern

Use the objects in your bag.
Make a pattern.
Draw it.

Show the same pattern with different objects.

# Same Number, Different Ways

Take one of your groups of objects.

Count the objects.

Arrange your group.

Draw it.

```


```

Arrange your group in a different way.

Draw it.

```


```

# All About Eleven

Take 11 objects from your bag.
Show different ways to build 11.

## Pattern Hunt

Look around your home.
Find some patterns.

Draw what you find.
Which room has the most patterns?

Design a new pattern for one of the rooms in
your home.

**Fold**

# Math at home

Math, math, everywhere
You say, "How can that be?"
At home, at school,
and all around,
Just look and you will see.

## At the Zoo

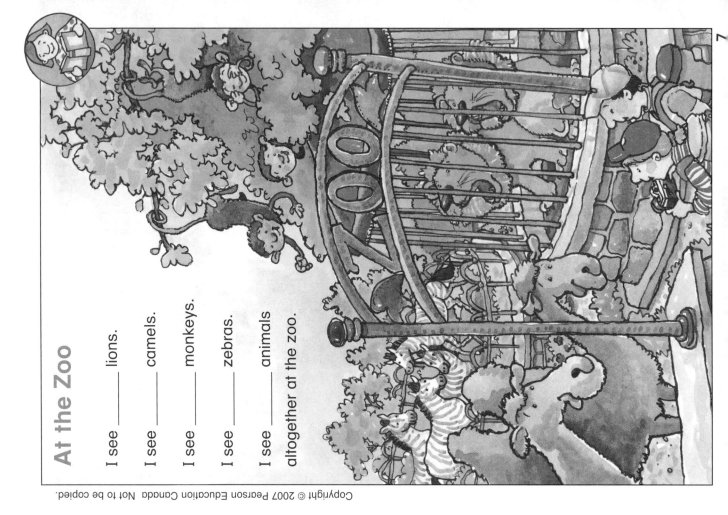

I see _____ lions.

I see _____ camels.

I see _____ monkeys.

I see _____ zebras.

I see _____ animals altogether at the zoo.

## Number Walk

Someone says a number between 2 and 18.
Everyone hunts for a number "2 more than" that.
The one who finds it first gets to pick a new number.
Now try finding "2 less than" numbers.

## Arrange It!

Ask a friend to pick a number between 2 and 12.
Use counters to build it in as many ways as you can.
Tell about your ideas.

## Setting the Table

How many of each do you need?

Is there a chair for every person?
Is there a knife for every fork?
Is there a plate for every glass?

What if someone else comes to dinner?
How many then?

# Under Cover

**Game**

**You'll need:**
- 12 small objects (pennies, buttons, bingo chips)
- 2 pieces of paper

**On your turn:**
- Ask your partner to turn around.
- Place some objects on one piece of paper in an interesting way.

8 could look like

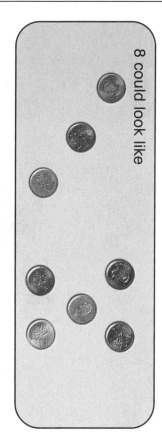

- Have your partner guess how many you used.
- Uncover your objects for 3 or 4 seconds. Your partner can look now.
- Cover the objects with the other paper.

- If the guess is right, score 2 points.
- If the guess is close, score 1 point.
- Decide together how many points you need to win!

---

# What's My Rule?

What comes next?

Start a pattern.
Keep adding to it until your partner knows what comes next.

Now your partner can start a pattern.
It's your turn to tell what the pattern is.

**Try:** shapes, coins, pasta, buttons, Snap Cubes

# Dough Patterns

Use play dough or cookie dough.
Roll out the dough.
Make or press a pattern into the dough.
Tell about your idea.

## More or Less Game Board

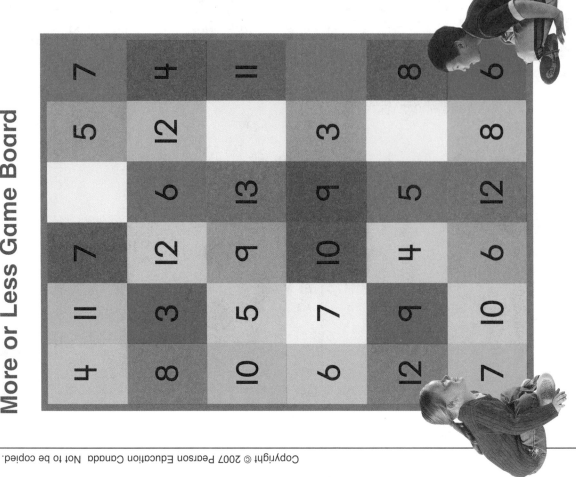

| 4 | 11 | 7 | | 5 | 7 |
|---|----|---|---|---|---|
| 8 | 3 | 12 | 6 | 12 | 4 |
| 10 | 5 | 9 | 13 | | 11 |
| | | 10 | 9 | 3 | |
| 9 | 7 | | 10 | | 8 |
| 12 | 9 | 4 | 5 | | 8 |
| 7 | 10 | 6 | 12 | | 9 |

How could you make this a "1 less than" game?

Aren't numbers fun?

## More or Less · Game

### You'll need:
- buttons for you
- buttons for your partner
- 2 number cubes
- the More or Less Game Board

### On your turn:
- Roll 2 number cubes.
- Count all the dots.
- Say the number that is "1 more than" you counted.
- Find the "1 more than" number on the game board.
- Cover it with one of your buttons.

Now it's your partner's turn.

### Make Up Your Own Rules
What happens when you can't cover a number?
What will you do with the blank squares?
When is the game over?

# UNIT 3

# Addition and Subtraction to 12

**FOCUS**

Children identify the number of bowling pins that are up and down in each lane and practise building number combinations of 5.

**HOME CONNECTION**

Have your child describe how each lane shows a way to make 5.

Name: _____   Date: _____

Dear Family,

In this unit, your child will be learning about addition and subtraction to 12.

The Learning Goals for this unit are to

- Learn more about the meaning of numbers.
- Recognize that addition involves joining groups and that subtraction involves taking one group away from another.
- Use counters and other real-life materials to help solve simple number problems.
- Practise using addition and subtraction to write and solve number problems.
- Describe the thinking involved in solving simple number problems.
- Use mental-math strategies to add and subtract.

You can help your child reach these goals by doing the activities suggested at the bottom of each page.

Name: _____ Date: _____

# What Is in the Box?

There are 10 marbles in a box.
Some are green. Some are yellow.
There are 2 more yellow marbles than green marbles.

How many yellow marbles are there?
Show your thinking in pictures, numbers, or words.

There are _____ yellow marbles.

 **FOCUS**

Children figure out a combination of green and
yellow marbles that makes 10, using a given
condition. They show their solution in pictures,
numbers, or words.

 **HOME CONNECTION**

Ask your child to describe how he or she solved the
problem.

# What Is in the Bag?

There are 11 balloons in a bag.

Some are red and some are purple.

There are 3 fewer purple balloons than red balloons.

How many purple balloons are there?

Show your thinking in pictures, numbers, or words.

There are _____ purple balloons.

 **FOCUS**

Children figure out a combination of red and purple
balloons that makes 11, using a given condition. They
show their solution in pictures, numbers, or words.

 **HOME CONNECTION**

Together, look at the solution your child recorded on
the page. Have your child explain the solution to the
problem.

# More Subtraction Stories

Use counters to make a subtraction story about children playing together.

Show your story using pictures, numbers, or words.

**FOCUS**

Children make a subtraction story using counters. They show their story in pictures, numbers, or words.

**HOME CONNECTION**

Have your child tell you the story on this page. Ask: "How did you think of that story? How did you decide how many to take away?"

# How Many Are Missing?

Look at the pictures. Record the numbers.

There are _____ ants

Now there are _____ ants.

_____ ants went inside.

_____ − _____ = _____

There are _____ bees

Now there are _____ bees.

_____ bees went inside.

_____ − _____ = _____

 **FOCUS**
Children build and identify subtraction stories
(missing part).

 **HOME CONNECTION**
Use 12 beads or buttons to play a hiding game. Hide
6 in your hand, leaving 6 visible. Ask: "How many
are missing?" Repeat, taking turns.

Name: _____ Date: _____

# Missing Counters

How many counters does each hand cover?
Record the numbers.

How many are covered? _____

How many are covered? _____

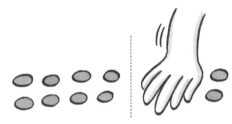

How many are covered? _____

How many are covered? _____

How many are covered? _____

How many are covered? _____

Name: _____ Date: _____

# Mental Math

Roll 2 number cubes.

What numbers did you roll?

Use mental math to add the numbers.

_____ **+** _____ **=** _____

Show your thinking in pictures, numbers, or words.

 **FOCUS**

Children use mental math to add 2 numbers.

**HOME CONNECTION**

Use the 2 numbers on this page and ask your child to use mental math to subtract the lesser number from the greater number. Ask your child to explain how he or she got the answer.

Name: _____  Date: _____

# Domino Dash

Think of the domino you picked from the bag.

What pictures did you make in your head?

What strategy did you use to add?

Show your thinking in pictures, numbers, or words.

_____ + _____ = _____

**FOCUS**
Children pick a domino from a bag and add the dots
using mental-math strategies.

**HOME CONNECTION**
Fold a piece of paper in half. Gather 12 pennies or
other small objects. Place up to 6 pennies on each
side of the paper. For example, 4 pennies and 2
pennies. Ask you child to use mental math to add
the pennies. Repeat with different arrangements
of pennies.

**Unit 3, Lesson 7:** Mental Math

# Add or Subtract

Write each number sentence.

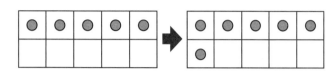

$$\underline{5} \bigcirc \underline{\quad} = \underline{6}$$

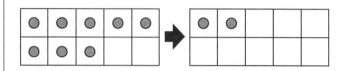

$$\underline{8} \bigcirc \underline{\quad} = \underline{2}$$

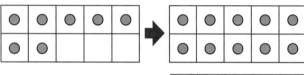

$$\underline{7} \bigcirc \underline{\quad} = \underline{12}$$

$$\underline{10} \bigcirc \underline{\quad} = \underline{3}$$

( Make Your Own ) Draw counters. Write a number sentence.

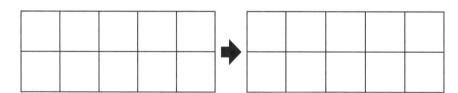

$$\underline{\quad} \bigcirc \underline{\quad} = \underline{\quad}$$

 **FOCUS**

Children interpret, build, and record addition and subtraction sentences using ten frames. Children create their own addition or subtraction sentence using a ten frame.

 **HOME CONNECTION**

Together, look at the sentence that your child built and recorded. Ask: "How did you think of that sentence? How did you decide how many to add or take away?"

Name: _____ Date: _____

# Make a Number Story

Draw counters to make a number story.
Write a number sentence.

11

3 fly away.

_____ ◯ _____ = _____

7

5 more come.

_____ ◯ _____ = _____

5

5 jump in.

_____ ◯ _____ = _____

6

6 fly away.

_____ ◯ _____ = _____

 **FOCUS**

Children draw counters to make number stories.
Then they record a number sentence for each
number story.

 **HOME CONNECTION**

Ask your child to tell you each number story and read
its number sentence.

**Unit 3, Lesson 8:** Combining and Separating Stories **81**

Name: _____ Date: _____

# My Story of 12

Use pictures, numbers, or words to tell a story of 12. You can use counters to help you.

My number sentence is _____ ◯ _____ = _____.

 **FOCUS**

Children use pictures, words, and numbers to tell a number story about 12. Then they write a number sentence for their story.

 **HOME CONNECTION**

Ask your child to tell you the story and read the number sentence. Ask: "Can you think of another number story for 12?"

Name: _____ Date: _____

# On and Off the Bus!

Write a number sentence to tell what happens each time.

3 children are on the bus.
At the next stop, 5 more get on.

_____ ◯ _____ = _____
_____

At the next stop, 4 children get off the bus.

_____ ◯ _____ = _____
_____

At the next stop, I child gets off the bus.

_____ ◯ _____ = _____
_____

At the last stop, 3 children get off the bus.

_____ ◯ _____ = _____
_____

How many children are on the bus now? _____

---

 **FOCUS**

Children use counters to model a story: Children are on a bus; each time the bus stops, some children get on or off. Children write number sentences to tell what happens.

 **HOME CONNECTION**

Ask your child to retell the story on this page. Use small objects to keep track of children getting on and off the bus. Have your child read the number sentences that help to tell the story.

**Unit 3, Lesson 9:** Show What You Know

Name: _____  Date: _____

# My Journal

Why are adding and subtracting important?
Use pictures, numbers, or words to show your thinking.

**FOCUS**
Children reflect on and record what they learned
about addition and subtraction in this unit.

**HOME CONNECTION**
Ask your child: "What do you like best about adding
and subtracting? Why do you think addition and
subtraction are important?"

# Measurement

 **FOCUS**

Children talk about comparing objects at home, at school, and in the community.

 **HOME CONNECTION**

Invite your child to describe the pictures. Ask: "What are these people doing? What do they want to find out? Why is it important to compare these things?"

Copyright © 2007 Pearson Education Canada Not to be copied.

**85**

Dear Family,

In this unit, your child will be learning about measurement.

The Learning Goals for this unit are to

- Make comparisons between objects (longer, taller).
- Order lengths of objects (shortest to longest).
- Compare the areas of objects by covering them with different materials.
- Practise the language of measurement using words such as *longer/shorter, as tall as, more/less, heavier/lighter.*
- Compare the capacities of containers by filling them with water and other materials.
- Compare the masses of objects using simple scales and balances.

You can help your child reach these goals by doing the activities suggested at the bottom of each page.

Name: _____ Date: _____

# I Can Compare

Show what you are comparing.

I compared _____.

I found out _____

_____.

 **FOCUS**

Children draw pictures of objects to compare, sharing what they compared and what they found out.

 **HOME CONNECTION**

Have your child describe the picture. Talk about what comparisons are made at home, such as heights on a growth chart.

Name: _____ Date: _____

# Longer or Shorter?

Which object is longer? Which object is shorter?

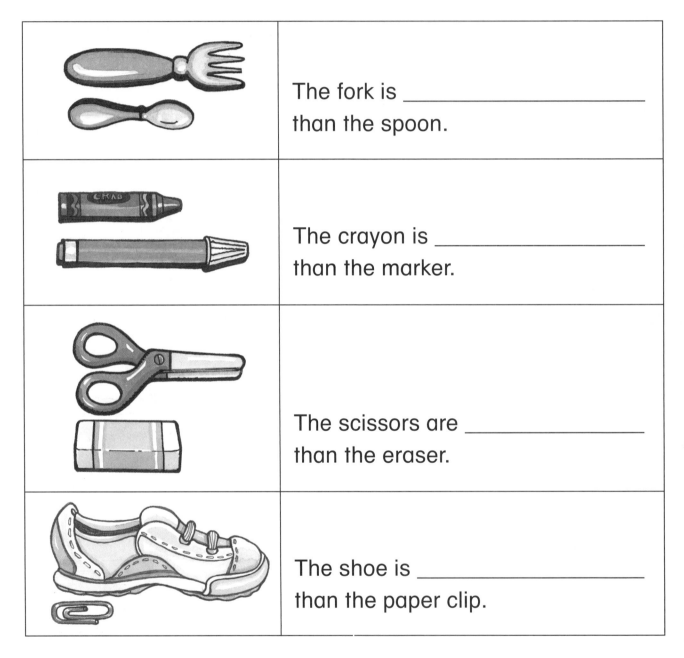

|  |  |
| --- | --- |
|  | The fork is _____ than the spoon. |
|  | The crayon is _____ than the marker. |
|  | The scissors are _____ than the eraser. |
|  | The shoe is _____ than the paper clip. |

Name: _____ Date: _____

# Object Hunt!

Use 5 Snap Cubes to make a train.

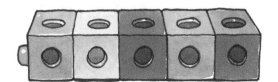

Draw an object that is shorter than the train.

The _____ is shorter.

Draw an object that is longer than the train.

The _____ is longer.

Draw an object that is about as long as the train.

The _____ is about as long.

**HOME CONNECTION**
Choose an object, like a spoon, and find objects that are shorter than, longer than, and about the same length. Have your child make comparison statements about the objects (e.g., The vase is longer than the spoon.).

# Ordering Lengths

Put the strings in order.

I put the strings in order from _____

to _____.

Think of another way to order the strings.

# Who Went the Farthest?

The _____ went the farthest distance.

The _____ went the shortest distance.

**FOCUS**

Children compare distances to determine the animal that went the farthest.

**HOME CONNECTION**

Cut a short length of string. With your child, find three things that are curved. Use the string to compare which is the longest and which is the shortest.

# Which Way Is Shorter?

Colour the shorter way to the ramp. Tell a partner how you know.

 **FOCUS**

Children identify the shorter route.

 **HOME CONNECTION**

With your child, order a collection of household items (such as cans or cereal boxes) from tallest to shortest.

# Compare Objects

Circle your answer.

I think (**my desk / my chair**) has the greater surface.

I needed (**more / less**) cards to cover my desk than the seat of my chair.

My desk has a (**greater / lesser**) surface than my chair.

Show another way to compare your desk to the seat of your chair.

---

 **FOCUS**
Children show ways to compare the surface of a desk to the surface of the seat of a chair.

 **HOME CONNECTION**
With your child, use a sheet of newspaper to cover a flat surface. Ask: "Which surface has the greater area?" Find other flat surfaces to compare.

Name: _____     Date: _____

# Cover to Compare!

Compare 3 objects.
Use a piece of paper.

| | | |
|---|---|---|
| 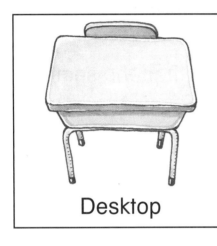  Desktop | Calendar | My Object _____ |

The _____ has the greatest area.

The _____ has the least area.

Use pictures or words to explain your thinking.

Name: _____ Date: _____

# Heavier or Lighter?

Which object is heavier? Which object is lighter?
Use real objects to find out.

The pencil is _____
than the book.

The stapler is _____
than the Pattern Block.

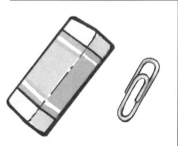

The eraser is _____
than the paper clip.

The chalk brush is _____
than the marker.

 **FOCUS**
Children compare some common objects to
determine which is heavier or lighter.

 **HOME CONNECTION**
Collect several household objects. Have your child
practise holding one in each hand to determine
which one is heavier and which one is lighter.

Unit 4, Lesson 6: Comparing Masses **99**

Name: _____   Date: _____

# Lighter, About the Same, or Heavier?

Look at each picture.

Is the object **lighter than, about the same as,** or **heavier than** your notebook?

Circle your answer.

lighter   about the same   heavier

lighter   about the same   heavier

lighter   about the same   heavier

lighter   about the same   heavier

lighter   about the same   heavier

lighter   about the same   heavier

Name: _____ Date: _____

# The Queen's Display

What objects can you find for the Queen? Draw them.

| Longest | Heaviest |
|---|---|
| | |
| **Covers the Greatest Area** | **Holds the Most** |
| | |

 **FOCUS**

Children find the biggest objects for the Queen of Giant's display by comparing.

 **HOME CONNECTION**

Ask your child: "Which objects at home would you choose for the Queen of Giant's display? Tell me about your thinking."

Name: _____ Date: _____

# My Journal

Tell what you learned about comparing lengths, areas, masses, and comparing by filling. Use pictures or words.

**FOCUS**

Children reflect on and record what they learned about measurement.

**HOME CONNECTION**

Ask your child to describe what he or she liked best about comparing lengths, areas, masses, and comparing by filling.

# Visiting the Fire Hall

Mr. Gloshes took the class to meet
The firefighters down the street.
His sister Sue works at the hall.
She's going to show it to them all.

She shows the firefighters' suits.
She shows the helmets, coats, and boots.
She shows the red truck, shining bright.
With ladders, hose, and flashing light.

Suddenly a message is phoned through.

A man shouts, "Sue, this one's for you."

"I have to go," Sue tells them all.

"We have an emergency rescue call."

"There is a cat stuck in a tree.
It's a very special cat, you see.
We won't be long. It isn't far.
So wait for us. Stay where you are."

The children gathered at the door.
Some children even paced the floor.
Then Sue came back! She held her hat.
In that helmet sat a cat!

Mr. Gloshes' eyes were wide.

"That's my cat!" Mr. Gloshes cried.

"Thank you for bringing him back to me.

I don't know how he climbed that tree."

"You're welcome," said his sister Sue.
"Helping out is what we do."
"Give a cheer—hip, hip, hooray!
Firefighters save the day!"

## From the Library

Ask the librarian about other books to share
about numbers, collecting, comparing and ordering objects,
and mass and capacity.

Suzanne Aker, *What Comes in 2's, 3's, and 4's?*
  (Aladdin Library, 1992)

Pamela Allen, *Who Sank the Boat?* (Puffin, 1996)

Aubrey Davis, *The Enormous Potato* (Kids Can Press, 2003)

Pamela Duncan Edwards, *Roar! A Noisy Counting Book*
  (HarperCollins Canada, 2000)

Simon James, *Dear Mr. Blueberry* (Aladdin Library, 1996)

Laura Joffe Numeroff, *If You Give a Moose a Muffin*
  (Scott Foresman, 1991)

Ann Tompert, *Just a Little Bit* (Houghton Mifflin, 1996)

# At the Fire Hall

Make number stories about the picture.

# Fire Hall Number Stories

Use your counters to make
an addition story about the fire hall.
Use pictures, numbers, or words.
Write an addition sentence for your story.

My story




My addition sentence _____

Make a subtraction story, too.

My story




My subtraction sentence _____

Hint Count the objects and people in the picture.

# Firefighters, More or Less

Count! How many firefighters? _____

2 more come. How many now?

I gets called away. How many now?

2 leave for lunch. How many now?

I comes back. How many now?

# Rescue Towers

Build a tower.
Use 5 objects.
My Tower:

Talk about how tall your tower is.

Compare your tower to the teacher's tower.

Is it taller, shorter, or the same?

Tell how you know. Use pictures and words.

# How Much Will It Hold?

Collect lots of different-sized containers from around your home.

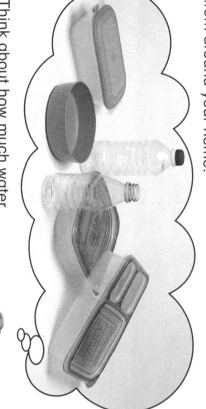

Think about how much water each container might hold, and order the containers from the smallest to the largest.

Now add water!
It's time to see if you were right.

## What Number Am I?

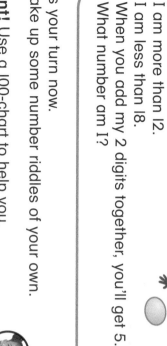

I have 2 digits.
I am more than 12.
I am less than 18.
When you add my 2 digits together, you'll get 5.
What number am I?

It's your turn now.
Make up some number riddles of your own.

**Hint!** Use a 100-chart to help you.

**Fold**

8

The next ll pages fold in half to make an 8-page booklet.

---

# Math at Home

Math is so much more, you see than numbers in a book.
It's measuring, building, games, and fun!
Let's take a closer look!

**Math at Home 2**

## Spinner

| 2 more | 2 less |
|--------|--------|
| 1 more | 1 less |

### Challenge

Write a number sentence to match what happened each time.

How many times did you have to spin to get exactly 12 toys?

---

## Doubles Hunt

Look in a mirror and search for all the parts of your body that come in groups of 2. How many can you see? Find out by counting by 2s!

## Balancing Act

What's wrong with this picture?

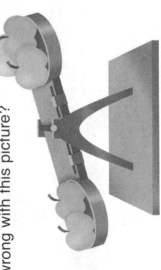

Draw items on either side so this picture makes sense. What is another way to balance the scale?

## More Toys, Please!

**Game**

### You'll need:

- a paper clip
- a pencil
- the spinner on page 7
- 12 toys for each player

### How to play:

- You can play alone or with a partner.
- Make a spinner with the paper clip and the pencil.
- Put the toys in a pile between you.
- Take turns to spin the paper clip.
- Try to get exactly 12 toys.

### On your turn:

- Take or give back the number of toys shown on the spinner.
- How many toys will you have now?
- Say the new number.

If you do not have enough toys to put back when the spinner lands on "1 less" or "2 less", you miss a turn.

## Comparing Shoes

Do you ever wonder about the lengths of shoes in your home?
Well, wonder no more!
It's time to find out!

- Gather all the shoes together and put them in a big pile.
- Compare the length of one of your shoes to the length of an adult shoe.
  Which shoe is longer?
  Are you surprised?

Find shoes that are shorter, longer, and about the same length as the length of your shoe.

# Parking Lot Game Boards

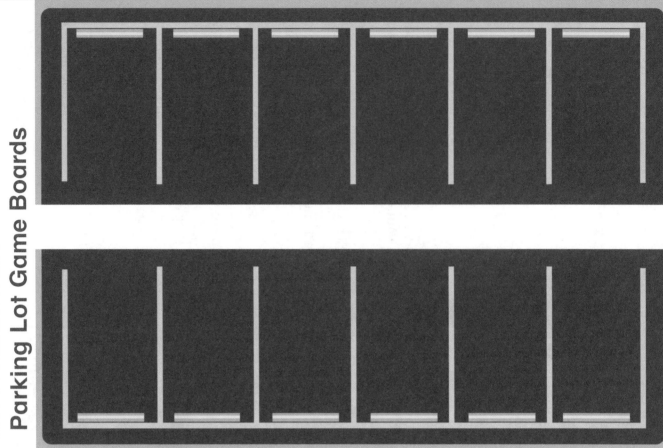

## Park-a-Lot Game

**Game**

### You'll need:
- 12 small toy cars (or any small objects)
- one number cube
- the parking lot game boards on the following page

### On your turn:
- Roll the number cube.
- Count the dots.
- Park the car in the matching spot. (If you roll a 2, you park your car in spot number 2.)

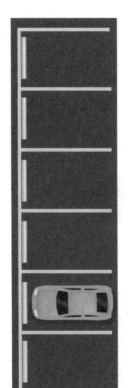

Now, it's your partner's turn.

If you roll a number and a car is already parked in that spot, miss your turn.

When your lot is full, the game is over.

Could you play this game with 12 parking spots?

How many number cubes would you need?

# Numbers to 100

 **HOME CONNECTION**

Look at this picture with your child. Ask your child: "How many children do you think there are at the concert?" Count the children together.

Name: _____   Date: _____

Dear Family,

In this unit, your child will be exploring some important mathematical concepts about numbers to 100.

The Learning Goals for this unit are to

- Read and record numbers to 100.
- Count sets in different ways.
- Group by 2s, 5s, and 10s.
- Count large collections of objects (up to 100).
- Skip count by 2s, 5s, and 10s.

You can help your child reach these goals by doing the activities suggested at the bottom of each page.

Name: _____ Date: _____

# Number Mystery

Fill in the missing numbers.

| 1 | 2 | 3 | 4 | | 6 | 7 | 8 | 9 | 10 |
|---|---|---|---|---|---|---|---|---|---|
| 11 | 12 | 13 | 14 | | | | 18 | 19 | 20 |
| 21 | 22 | | 24 | 25 | 26 | 27 | 28 | 29 | 30 |
| 31 | 32 | | 35 | 36 | 37 | 38 | 39 | | |
| | 42 | 43 | 44 | 45 | 46 | 47 | 48 | 49 | |

Share your chart with a partner.

Talk about how you found the missing numbers.

 **FOCUS**
Children fill in the missing numbers on the 50-chart.

 **HOME CONNECTION**
Show your child a page number (between 20 and 50) in a book or magazine. Ask: "What will the next page number be? What page number came before?" Repeat with different pages.

# Count Two Ways

Spill the objects.
Count the objects.

> Show how you counted. Use pictures, numbers, or words.

Spill the objects again. Count them another way.

> Show how you counted. Use pictures, numbers, or words.

 **FOCUS**

Children count a collection of objects in two different ways. They record their counting methods using pictures, numbers, or words.

 **HOME CONNECTION**

Gather up to 50 pennies or other small objects. Ask your child to show you two different ways of counting them.

Name: _____ Date: _____

# Bee Count

## How many bees are there?
## Show how you counted.

```
┌──────────────────────────────────────────────────────────────────┐
│                                                                    │
│                                                                    │
│                                                                    │
│                                                                    │
│                                                                    │
│                                                                    │
│                                                                    │
└──────────────────────────────────────────────────────────────────┘
```

## Count the bees another way.
## Show how you counted.

```
┌──────────────────────────────────────────────────────────────────┐
│                                                                    │
│                                                                    │
│                                                                    │
│                                                                    │
│                                                                    │
│                                                                    │
│                                                                    │
└──────────────────────────────────────────────────────────────────┘
```

 **FOCUS**
Children count the bees using different-size groups.

 **HOME CONNECTION**
Pose a problem to your child, such as: "How many knives, forks, and spoons are in the drawer? Think of two different ways to count them."

Date: _____

# w Many Counters?

ᴜke a scoop of counters.

Place the counters in the ten frames.

How many groups of 10? _____

How many in all? _____

**FOCUS**

Children scoop some counters. They place the
counters on the ten frames to count.

**HOME CONNECTION**

Place 30 to 50 small countable objects (beans,
paper clips) in a plastic bag. Ask: "How many
objects are in the bag?" Have your child count the
objects by placing them in groups of 10.

Name: _____ Date: _____

# Count the Beans

Draw 47 beans.
Circle groups of 10.

There are _____ groups of 10 and _____ left over.

Print the number.

 **FOCUS**
Children draw 47 beans and circle groups of 10.
They record the number of beans as a numeral.

 **HOME CONNECTION**
Select a number between 20 and 30. Ask your child
to draw the same number of objects. Help your child
circle groups of 10. Ask: "How many groups of 10?
How many left over?"

.ne: _____ Date: _____

# How Many Wheels?

We need wheels for 5 bicycles.

How many wheels should we get?

Show your thinking in pictures, numbers, or words.

 **FOCUS**

Children draw a picture to solve a problem. They express their solution using pictures, numbers, or words.

 **HOME CONNECTION**

Have your child explain how he or she solved this problem. Ask: "How did you know the number of wheels needed for 5 bicycles? Tell me about your thinking."

Name: _____ Date: _____

# How Many Bicycles?

A bicycle rack holds 5 bicycles.
The school has 4 bicycle racks.
How many bicycles can be parked?

Show your thinking in pictures, numbers, or words.

 **FOCUS**

Children make a model or use objects to solve a problem. They express their solutions in pictures, numbers, or words.

 **HOME CONNECTION**

Have your child explain how he or she solved this problem. Ask: "How did you know the number of bicycles? Tell me about your thinking."

Unit 5, Lesson 7: Strategies Toolkit

Name: _____  Date: _____

# How Many Buttons?

Look at the button blanket.

Circle groups of 10. Use an orange crayon.

How many groups of 10? _____

How many are left over? _____

How many in all? _____

Circle groups in another way. Use a green crayon.
Tell a friend how you counted.

 **FOCUS**
Children count the buttons on a button blanket using
10s and some left over, and in another way.

 **HOME CONNECTION**
Ask your child to explain how he or she solved the
problem. Ask: "How did you make sure you didn't
count any buttons twice?"

Name: _____  Date: _____

# My Button Blanket

How many groups did you use? _____ groups of _____

How many extra buttons? _____

How many buttons in all? _____

 **FOCUS**

Children design a button blanket. They record the
number of buttons used.

 **HOME CONNECTION**

Ask your child to explain how he or she decided how
many buttons to draw, and how to group them.

Name: _____  Date: _____

# My Journal

How can you group by 10s to help you count?
Use pictures, numbers, or words to show your thinking.

 **FOCUS**
Children reflect on and record what they learned
about numbers to 100 in this unit.

 **HOME CONNECTION**
Ask your child, "What did you learn about how to
count to 100? How can you group by 10s to help
you count?"

**138**　**Unit 5, Lesson 8:** Show What You Know

Copyright © 2007 Pearson Education Canada **Not to be copied.**

# Geometry

 **FOCUS**

Children find and describe 3-D objects in this picture.

 **HOME CONNECTION**

Ask your child to describe the objects that make up the castle.

Name: _____  Date: _____

Dear Family,

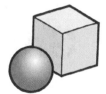

Your child will be learning more about three-dimensional objects and two-dimensional shapes.

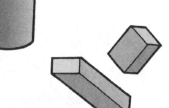

The Learning Goals for this unit are to

- Compare and sort objects or shapes according to one attribute, such as number of sides or whether they have curves.

- Describe objects and shapes. For example, some objects can roll, some have points, some are curved, and some can stack.

- Reproduce composite objects and shapes, such as towers.

- Identify shapes on objects, such as a circle on the bottom of a yogurt container.

You can help your child reach these goals by doing the activities suggested at the bottom of each page.

Name: _____ Date: _____

# Sorting Objects

Sort and paste pictures.

| These have only flat faces. |
| --- |
|  |

| These have curved faces. |
| --- |
|  |

 **FOCUS**

Children cut out pictures of objects from *Line Master I: Different Objects*. Then they sort the objects based on an attribute and paste them into a chart.

 **HOME CONNECTION**

Gather a collection of different objects. Ask your child to find all the objects that can stack.

Unit 6, Lesson I: Sorting 3-D Objects **143**

# Our Tall Tower

Draw a picture of your tower.

Show the objects you used to build your tower.

 **FOCUS**

Children record the objects they used to replicate a tower. Later, they will compare towers with other pairs of children.

 **HOME CONNECTION**

Have your child experiment by building tall towers out of blocks, small boxes, cans, or other household objects.

# I Am a Builder

What objects did you use for your structures?

| Objects | How Many Objects? |
|---|---|
|  |  |
|  |  |
|  |  |
|  |  |
|  |  |

How are the structures different?

Use pictures, numbers, or words.

**FOCUS**

Children record the objects they used to build two structures, then tell how the structures are different.

**HOME CONNECTION**

With your child, collect objects to use with this unit: party hats, cans, toilet-tissue rolls, cereal boxes, milk cartons, balls, and square boxes.

Name: _____    Date: _____

# My Shape

Tell about your shape.

[ ]

Compare with a friend's shape.
How are your shapes alike?

Both have _____.

How are they different?

One has _____. One does not.

 **FOCUS**

Children select one shape and tell about it. They compare shapes with a partner and note similarities and differences.

 **HOME CONNECTION**

Invite your child to tell about the shape described above. Ask: "What does it remind you of? How would you describe it to someone who cannot see it?"

Name: _____    Date: _____

# Sorting Shapes

| Circle the shapes that are alike. | Paste a shape that is alike. |
|---|---|
| 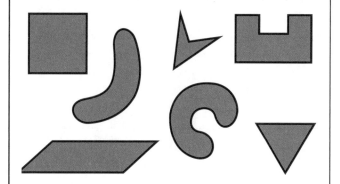 | |

It is alike because it has _____.

| Circle the shapes that are alike. | Paste a shape that is alike. |
|---|---|
| 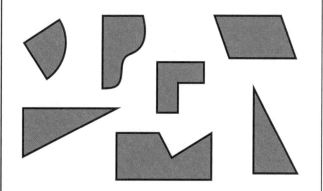 | |

It is alike because it has _____.

Name: _____  Date: _____

# Matching Shapes

| Object | My Matching Shape |
|---|---|
|  |  |
|  |  |
|  |  |
|  |  |
|  |  |

 **FOCUS**

Children draw matching shapes for each illustration.

 **HOME CONNECTION**

Choose a geometric attribute of a favourite toy—corners, straight sides, curves, dents—and ask your child to find other objects with the same attribute.

**148**  **Unit 6, Lesson 3:** Sorting 2-D Shapes

Name: _____ Date: _____

# Match the Tower

| Objects | My Tower |
|---------|----------|
|         |          |

Tell how you made your tower.

Use pictures, numbers, or words.

|  |
|--|

**FOCUS**

Children reproduce a tower from a picture and then tell how they made their tower.

**HOME CONNECTION**

Collect matching sets of household objects, such as a can, a paper towel roll, and a tissue box. Build a tower using the objects, then have your child replicate the tower.

Name: _____ Date: _____

# My Journal

What did you learn about objects?
Show your thinking.

What did you learn about shapes?
Show your thinking.

 **FOCUS**
Children reflect on what they learned about 3-D
objects and 2-D shapes in this unit.

 **HOME CONNECTION**
Ask your child to describe a favourite activity from the
unit. Ask: "Why is it your favourite?"

**156** **Unit 6, Lesson 7:** Show What You Know     Copyright © 2007 Pearson Education Canada **Not to be copied.**

# Addition and Subtraction to 20

 **FOCUS**

Children discuss the picture and tell number stories.

 **HOME CONNECTION**

Ask your child: "What number stories can you tell about this picture?"

Name: _____   Date: _____

Dear Family,

In this unit, your child will be learning more about addition and subtraction to 20.

The Learning Goals for this unit are to

- Explore addition and subtraction facts.
- Use mental math strategies.
- Create and solve story problems.
- Explore "doubles" up to 9 + 9.
- Solve simple addition and subtraction problems.

You can help your child reach these goals by doing the activities suggested at the bottom of each page.

# Number Sentences

Write each subtraction sentence.

20 children have  .

7 children take them off.

_____ − _____ = _____

17  are on a shelf.

6 fall off.

_____ − _____ = _____

13  are standing by the fence.

7 fall over.

_____ − _____ = _____

14 are in a box.

6 get taken out.

_____ − _____ = _____

---

 **FOCUS**

Children write subtraction sentences about the pictures shown.

 **HOME CONNECTION**

Have your child tell about one of the number sentences recorded on the page. Ask: "How did you think of that number sentence? How did you know how many to take away?"

Name: _____ Date: _____

# Numbers Race!

Work with a partner.

Take turns. Find each missing number.

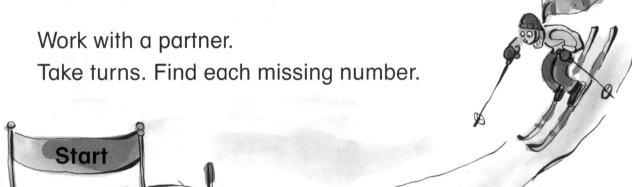

**Start**

$12 - 9 =$ _____

_____ $+ 4 = 10$     $4 +$ _____ $= 9$

$16 - 10 =$ _____

$8 +$ _____ $= 14$

_____ $+ 7 = 11$     $15 -$ _____ $= 9$

$15 +$ _____ $= 17$

$18 - 17 =$ _____

$13 - 8 =$ _____

_____ $+ 9 = 14$

$11 - 7 =$ _____

**Finish**

 **FOCUS**

Children work together to complete addition and subtraction sentences.

 **HOME CONNECTION**

Make up simple number stories for your child to solve. (For example: There are 8 shoes by the door, and 4 boots. How many shoes and boots are there altogether?)

# Count On, Count Back

$5 + 3 =$ _____   $9 + 2 =$ _____   $11 + 1 =$ _____   $14 + 2 =$ _____

$12 + 3 =$ _____   $9 + 11 =$ _____   $3 + 8 =$ _____   $2 + 15 =$ _____

Count back to subtract.

$10 - 3 =$ _____   $12 - 2 =$ _____   $14 - 3 =$ _____   $15 - 6 =$ _____

$20 - 2 =$ _____   $16 - 1 =$ _____   $13 - 2 =$ _____   $17 - 3 =$ _____

Write a number sentence.

12

3 fly away.

15

3 more come.

_____ ◯ _____ = _____      _____ ◯ _____ = _____

 **FOCUS**

Children use counting on and counting back to complete addition and subtraction sentences.

**HOME CONNECTION**

Have your child tell about one of the number sentences recorded on the page. Ask your child: "How did counting on or counting back help you find number sentences?"

**Unit 7, Lesson 4:** Strategies for Addition and Subtraction  **165**

# Relating Addition and Subtraction

Use the addition fact to find the related subtraction fact.

$7 + 3 = 10$  $\qquad$  $10 - 3 =$ _____

$5 + 6 = 11$  $\qquad$  $11 - 6 =$ _____

$8 + 9 = 17$  $\qquad$  $17 - 9 =$ _____

$9 + 6 = 15$  $\qquad$  $15 - 6 =$ _____

Use the subtraction fact to find the related addition fact.

$16 - 9 = 7$  $\qquad$  $7 + 9 =$ _____

$15 - 8 = 7$  $\qquad$  $7 + 8 =$ _____

$18 - 9 = 9$  $\qquad$  $9 + 9 =$ _____

$13 - 7 = 6$  $\qquad$  $6 + 7 =$ _____

 **FOCUS**
Children use a given addition or subtraction fact to write the related subtraction or addition fact.

 **HOME CONNECTION**
Ask your child: "How did you find the related subtraction or addition facts on this page? How else could you find the subtraction or addition facts?"

Name: _____  Date: _____

# By the River

What's happening by the river?

Tell an addition story.

Write the addition sentence.

_____ ◯ _____ = _____

Tell a subtraction story.

Write the subtraction sentence.

_____ ◯ _____ = _____

**FOCUS**

Children tell addition and subtraction stories about animals by the river.

**HOME CONNECTION**

With your child, pose story problems about things you and your child see or do. For example: "There are 9 juice boxes on the shelf. There are 3 juice boxes on the counter. How many juice boxes are there in all?"

Name: _____    Date: _____

# More Problems to Solve

Circle  **+**  or  **−**  to add or subtract.
Complete the number sentence.

There are 16 plants. 8 have flowers.
How many do not have flowers?

**+**          **−**

_____ ◯ _____ = _____

6 pink roses are in a vase. 6 red roses are added.
How many are there altogether?

**+**          **−**

_____ ◯ _____ = _____

14 leaves are on a branch. 5 fall off.
How many leaves are still on the branch?

**+**          **−**

_____ ◯ _____ = _____

There are 8 black ants. 10 red ants came along.
How many ants are there altogether?

**+**          **−**

_____ ◯ _____ = _____

 **FOCUS**
Children decide which operation to use to solve a
word problem, then solve the problem.

 **HOME CONNECTION**
Review the problems on this page with your child.
Together, make up similar problems and talk about
whether to add or subtract to solve them.

# Be a Problem Solver

Solve each problem.

There are 4  . There are 7  .
How many kittens in all?

There are _____  in all.

19  are in a pen. 6  run away.

How many  are left in the pen?

_____ rabbits are left in the pen.

Tell a number story to go with the picture. Write a number sentence.

_____ ◯ _____ = _____

**FOCUS**
Children use pictures, numbers, or words to solve addition and subtraction story problems.

**HOME CONNECTION**
Share addition and subtraction story problems using things you see in your neighbourhood. For example, "14 cars are parked on the street. 3 cars are red. How many cars are not red?"

Name: _____  Date: _____

# Choose Your Own Answer!

And the answer is _____!

Find as many ways as you can to get your answer.

**FOCUS**

Children find as many ways as they can to get the target number they select. They can both add and subtract.

**HOME CONNECTION**

The next time you go grocery shopping with your child, ask your child to add the number of items in your grocery cart.

# Neighbourhood Party

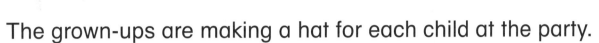

There is a party on Oak Street.

12 girls are coming. 8 boys are coming.

The grown-ups are making a hat for each child at the party.

How many hats should they make?

They should make _____ party hats.

The grown-ups are blowing up a balloon for each child.

They blow up 15 balloons.

How many more balloons do they need?

They need _____ more balloons.

 **FOCUS**

Children solve number-story problems to show what they learned about addition and subtraction.

 **HOME CONNECTION**

These problems give your child a chance to show what he or she learned about addition and subtraction to 20. Have your child explain the solution to each problem.

Name: _____  Date: _____

# My Journal

Tell what you learned about addition and
subtraction to 20.
Use pictures, numbers, or words.

---

 **FOCUS**

Children reflect on and record what they learned
about addition and subtraction in this unit.

**HOME CONNECTION**

Ask your child: "What did you learn about addition and
subtraction to 20?"

# Seeds, Seeds, Seeds

Mr. Bumbles scooped up some seeds.
"We'll count how many each group needs.
Then we'll plant them, one by one, and
put the plant pots in the sun."

They watered the seeds, and watched them too,
But still no plants were breaking through.
On Friday when it was time to go, the class
wondered, "On Monday, will plants show?"

They raced to look when Monday came.
Something in the room was not the same.
All they could do was stop and stare.
The plants were gone! The plants weren't there!

"How will we know if they've started to sprout?
Where did they go?" the class cried out.
Mr. Bumbles scratched his head.
"I think I know," Mr. Bumbles said.

Then someone knocked at the classroom door.
She knocked again, then knocked some more.
The school helper, Mary, was in the hall—
with every plant! She had them all!

"I took them home on Friday night.
I thought they might get too much light.
I watered them carefully every day.
I'm pretty sure they're all okay."